in my sticker era

this book belongs to

____

if found, please return... and bring coffee

notes

Lot No.
73503
82
PIECE NO.

notes

notes

notes

notes

the PCS
stickers
have
squatters'
rights

notes

notes

notes

notes

Lot No.
8675309
67
PIECE NO.

notes

notes

notes

notes

i didn't choose the sticker life, it chose me

notes

notes

notes

notes

Lot No.
002319
105
PIECE NO.

notes

notes

notes

notes

PCS
stickers
multiply
when no
one is
watching

notes

notes

notes

notes

notes

notes

notes

notes

i wish my
patience
had the
same grip
as my
stickers

notes

notes

notes

notes

Lot No.
95247
47
PIECE NO.

notes

notes

notes

notes

they are
my
emotional
support
stickers

notes

notes

notes

notes

notes

notes

notes

notes

Murphy's Law is fueled by PCS stickers

notes

notes

notes

notes

notes

notes

notes

notes

i'll peel
them next
duty
station

notes

notes

notes

notes

notes

notes

notes

notes

the
stickers
made it,
the
furniture
didn't

notes

notes

notes

notes

notes

notes

notes

notes

PCS once,
stick
forever

notes

notes

notes

notes

i tried to
move on,
the
stickers
followed

notes

notes

to-do lists

to-do

to-do

to-do

to-do

to-do

to-do

to-do

to-do

to-do

to-do

to-do

to-do

to-do

to-do

to-do

to-do

to-do

to-do

to-do

to-do

to-do

to-do

to-do

to-do

to-do

to-do

to-do

to-do

to-do

to-do

to-do

to-do

to-do

to-do

to-do

to-do

to-do

to-do

to-do

to-do

to-do

to-do

to-do

to-do

to-do

to-do

to-do

to-do

to-do

to-do

to-do

to-do

to-do

to-do

to-do

to-do

to-do

to-do

to-do

to-do

to-do

to-do

to-do

to-do

to-do

to-do

to-do

to-do

to-do

Joshua 1:9

CHECK OUT THE DANDELION KIDS CLUB COLLECTION!

Available on

amazon

If this notebook made your day a little easier or brought a smile to your face, please **consider leaving a review.**

It helps more than you know.

Thank you for supporting a
military spouse-owned small business!

thank you!

Joanna Gosser

Made in the USA
Monee, IL
30 July 2026

58510670R00090